discovermore

Exploring Physical Science

Friction

Marie Harts

Britannica Educational Publishing

IN ASSOCIATION WITH

THE ROSEN PUBLISHING GROUP

Published in 2024 by The Rosen Publishing Group, Inc.
2544 Clinton Street, Buffalo, NY 14224

Editor: Shannon Harts
Book Design: Rachel Rising

Photo Credits: Cover Dovzhykov Andriy/Shutterstock.com; (series background) Dai Yim/Shutterstock.com; p. 4 Sergey Novikov/Shutterstock.com; p. 5 LeManna/Shutterstock.com; p. 7 ChWeiss/Shutterstock.com; p. 7 luchschenF/Shutterstock.com; p. 8 Krakenimages.com/Shutterstock.com; p. 9 Zakirov Aleksey/Shutterstock.com; p. 11 yanik88/Shutterstock.com; p. 11 OlgaOvcharenko/Shutterstock.com; p. 13 asharkyu/Shutterstock.com; p. 13 Krasula/Shutterstock.com; p. 14 abimages/Shutterstock.com; p. 15 Prostock-studio/Shutterstock.com; p. 16 Reshetnikov_art. Shutterstock.com; p. 17 Designua/Shutterstock.com; p. 18 Sergey Zaykov/Shutterstock.com; p. 19 grayjay/Shutterstock.com; p. 20 Esther Pueyo/Shutterstock.com; p. 21 jamesteohart/Shutterstock.com; p. 23 Vagabondivan/Shutterstock.com; p. 23 Igor Link/Shutterstock.com; p. 24 Sibons photography/Shutterstock.com; p. 25 J nel/Shutterstock.com; p. 26 Galashevsky Yakow/Shutterstock.com; p. 27 SIMON SHIM/Shutterstock.com; p. 28 Krasula/Shutterstock.com; p. 29 Dima Zel/Shutterstock.com.

Cataloging-in-Publication Data

Names: Harts, Marie.
Title: Friction / Marie Harts.
Description: New York : Britannica Educational Publishing, in Association with Rosen Educational Services. 2024. | Series: Discover more: exploring physical science | Includes glossary and index.
Identifiers: ISBN 9781641900485 (library bound) | ISBN 9781641900478 (pbk) | ISBN 9781641900492 (ebook)
Subjects: LCSH: Friction--Juvenile literature.
Classification: LCC QC197.H37 2024 | DDC 531.1134--dc23

Manufactured in the United States of America

Some of the images in this book illustrate individuals who are models. The depictions do not imply actual situations or events.

CPSIA Compliance Information: Batch #CWBRIT24. For further information contact Rosen Publishing at 1-800-237-9932.

Contents

Defining Friction

Friction can be found in all sorts of places. It's a force (a push or pull) that acts between two objects that are touching one another. Friction slows or stops the movement between the surfaces of the objects that are touching.

Rolling creates less friction. This is why wheels on skates allow roller skaters to glide over hard surfaces like roads.

Friction on bike tires and a road allow you to roll safely without sliding off the road.

Friction can occur between all types of matter: solids, liquids, and gases. Sometimes it can be useful. At other times it can cause problems. Friction allows objects to stop, but it can also cause items such as the soles of shoes to be worn down. People experience both useful and problematic friction in their everyday lives.

Consider This

Friction can cause teeth to become worn down over time. How do you think this happens?

Time for a Close-Up!

How do scientists get a super close look at something's surface? They use a tool called a microscope! When they look at the surfaces of common objects, all sorts of tiny bumps, ridges, and pits show up. What about glass or slippery plastic? Looking carefully shows that this is true for even the smoothest of surfaces.

When objects touch or rub together, the bumps, ridges, and pits grip, grind, drag, and smash into each other. Rough surfaces, such as a carpeted floor, have more of these bumps and jagged edges. They produce a larger amount of friction. Although smooth surfaces also have bumps that can be seen under a microscope, they have fewer bumps that cause friction than rough surfaces. For example, although ice can look bumpy under a microscope, it's easier to glide over than rocks or sand.

This picture shows what ice looks like under a microscope. Even most surfaces that look flat and smooth, like ice, have tiny bumps that can cause friction.

compare and contrast

Think of a smooth surface, such as a piece of glass, and a rough surface, such as sandpaper. How do you think they would look similar and different under a microscope?

Surfaces that may look smooth, such as a leaf, look lumpy under a microscope because they are made up of atoms and molecules that don't line up in perfectly smooth ways.

Uses of Friction

No matter where you go, friction is involved in your everyday life. When people walk, friction between the soles of their shoes and the ground prevents them from slipping and sliding.

Since friction can create heat, if your hands are cold, you can rub them together quickly to warm them up!

Often rock climbers wear special shoes made of soft rubber that increases how much of the shoe touches the rock, and thus its friction with the rock. This helps keep the climber from slipping!

Imagine getting to school and not being able to draw, write, or color. Without friction, it would be impossible to grip your pencils, pens, or crayons, and you would not be able to mark your paper.

It may be fun and exciting to go fast on a bike, but eventually you will need to stop. You can stop thanks to the bike's brakes creating friction that slows and finally stops its wheels.

Consider This

Think about the many different designs and types of shoes you can wear, and how the soles are different. What do you think this has to do with friction?

Having Fun with Friction

You may not realize that your favorite sport or game depends on friction. Baseball is fun to play during the summer. A tip-top baseball player needs to grip the bat, hit the ball, and run rapidly. How about playing tag? It's exciting to run, turn, and stop quickly. These activities are possible with enough friction between your hands and the bat and between your shoes and the ground.

Swimming and sledding are fun too. Splashing down a slippery slide into a pool is delightful on a warm day, and zipping down a snowy hill in winter is equally thrilling. You can go fast down a slide or on a sled thanks to the smooth plastic sled and slide creating little friction. However, friction allows you to stop so you aren't sliding or sledding forever!

Have you played fetch with a dog? Dogs depend on their teeth creating friction to grab objects like frisbees.

compare and contrast

How is going down a waterslide different than going down a dry slide on a playground? Do you go faster down the waterslide?

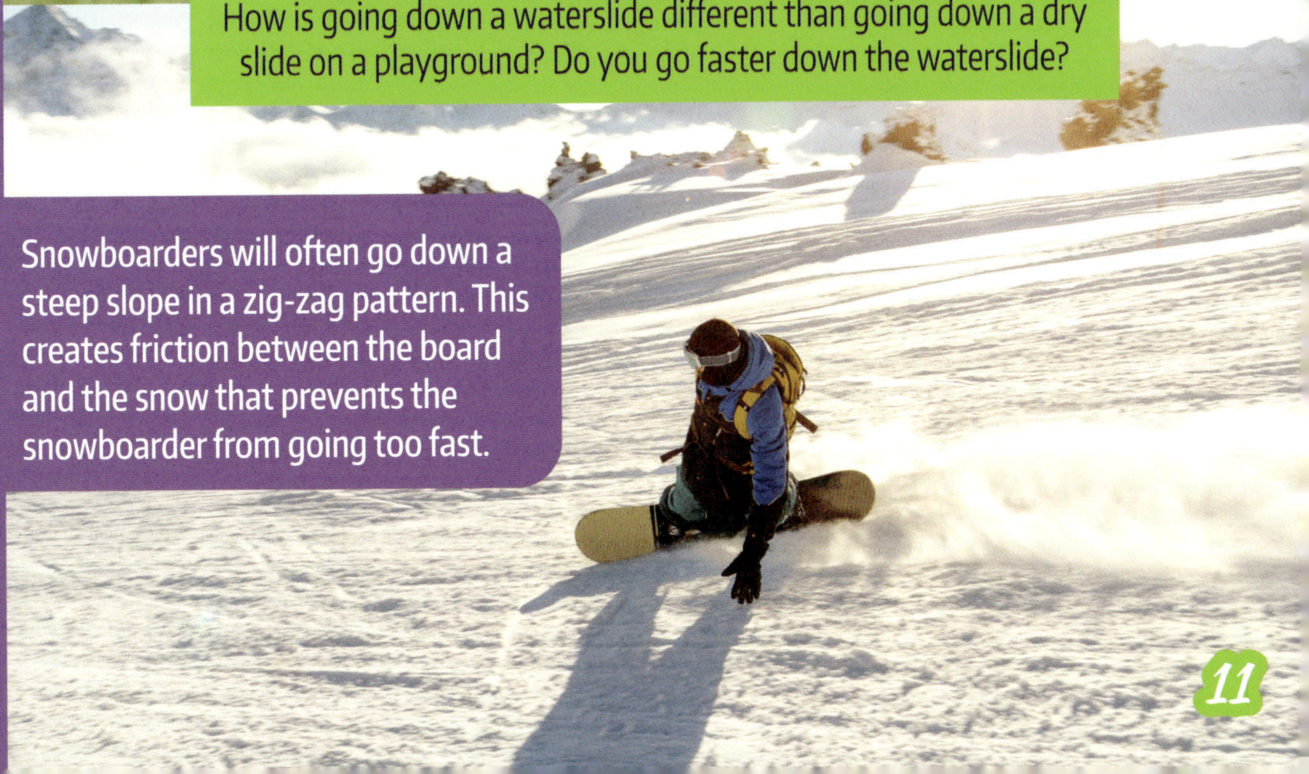

Snowboarders will often go down a steep slope in a zig-zag pattern. This creates friction between the board and the snow that prevents the snowboarder from going too fast.

Safety First!

Friction isn't just for fun—it can also keep you, your family, and your friends safe. During winter, when it's cold outside, streets and sidewalks may become icy and slippery. People try to melt the ice with salt. They also throw sand on top of it. This increases the friction between their shoes or boots and the icy surface. Better friction decreases the chances of slipping and falling.

In the shower, bathtub, and other wet areas, water decreases the friction between people's feet and the surface they are stepping on. This is why an adult has probably told you not to run near a pool! Rubber mats can provide friction to prevent people from slipping and falling near pools, tubs, and showers.

During and after snowstorms, plows spread salt on roads. Salt melts ice, increasing the road's friction to prevent car accidents.

compare and contrast

Covering an icy surface with sand can immediately increase its friction. Salt melts ice so the original surface's friction prevents slipping. Which do you think is safer?

Signs are often put out to warn people of wet floors that might cause them to slip and fall. This is because the liquid causes there to be less friction between the floor and people's shoes.

Friction's Static Form

Whenever two objects come into contact, there is friction—but not all friction is the same. If there's no movement between the objects, it's known as **static** friction.

To understand static friction, think about someone who is cleaning their bedroom. In the middle of the room is a large box filled with toys.

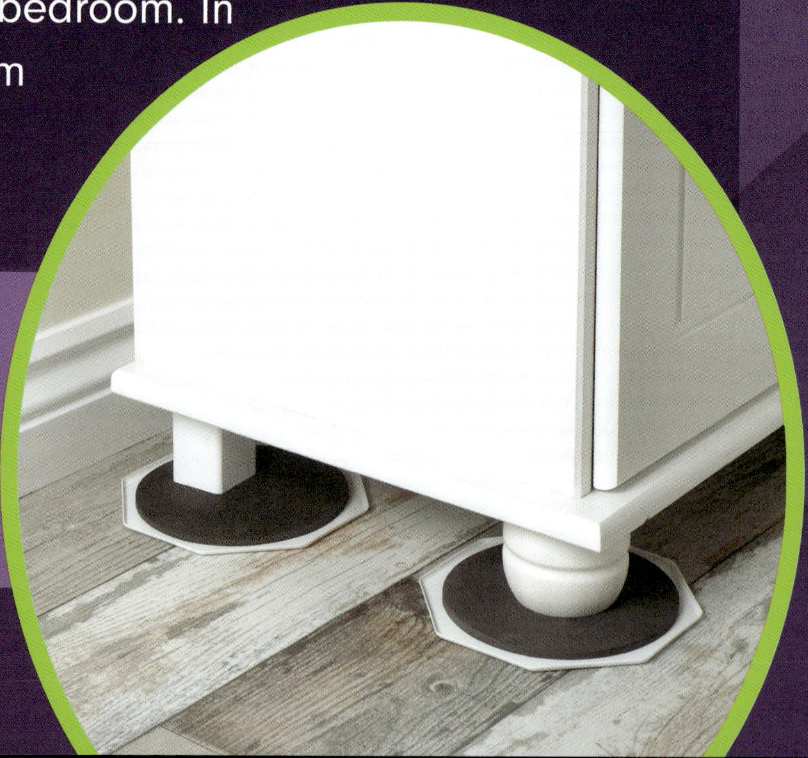

Round, flat pads called furniture sliders can be put under heavy furniture to reduce friction, making it easier to move.

Static friction allows glasses to stay on your nose! Without this form of friction, the glasses would slide off your face.

To finish the job, the box needs to be pushed into the closet. But it won't be easy. Static friction between the floor and the bottom of the box holds it in place. If the box gets heavier, the amount of static friction increases. This means that it will take more effort to overcome the static friction and move the box. Have you experienced this when trying to move heavy items?

WORD WISE

Static is a word to describe things that don't move or change.

Friction's Kinetic Form

If static friction happens when two objects are in contact and aren't moving, what is it called when two objects in contact are moving? It's **kinetic** friction! Kinetic friction happens when two things are moving or sliding against one another. Kinetic friction resists an object that is moving or sliding, eventually bringing it to a stop.

Kinetic friction can be found in ballet. It's a sliding motion that acts between a dancer's feet and the surface they are dancing on. Kinetic friction can help dancers keep their balance.

Sliding Friction

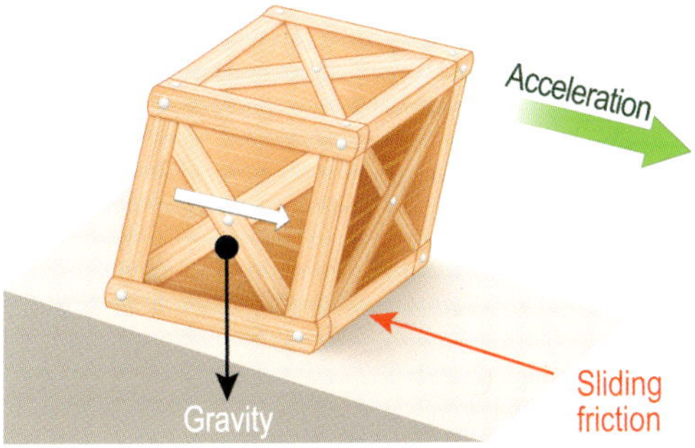

Rolling Friction

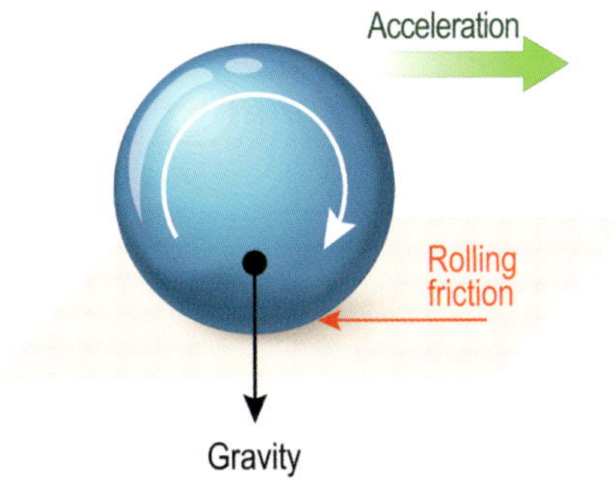

Kinetic friction comes in two forms: sliding friction and rolling friction.

Think back to the person with the large, toy-filled box. Once the person gets the box moving, it's not as difficult, or hard, to push and slide it along the floor. Kinetic friction makes it easier to continue moving the box than it was to get the box moving.

Consider This

Rolling friction involves a ball or wheel rolling over a surface while sliding friction involves two objects simply moving or sliding against each other. Can you think of examples of both types of kinetic friction in your life?

WORD WISE

Kinetic is a word used to describe objects that produce or involve movement.

Measuring Friction

When the surfaces of two objects meet, a certain amount of friction will be produced. With some substances, more friction occurs. With others, less friction occurs. The amount of friction created by different surfaces meeting can be measured. This measurement is called the **coefficient** of friction.

Ice skating is an example of using two objects with a low coefficient of friction: ice and steel. Friction allows skaters to stop, though!

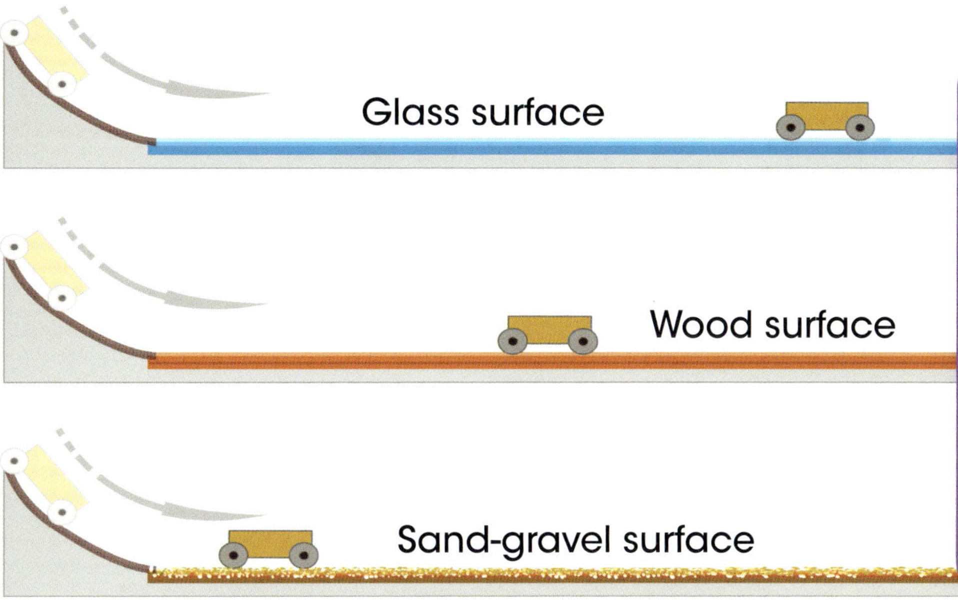

You can create an experiment using a toy car or other wheeled object to test different surfaces' coefficients of friction.

Surfaces that slow down movement the most are high friction. Sliding a concrete brick over a concrete road would be terribly difficult. So concrete is said to have a high coefficient of friction. Surfaces that allow more movement to occur have lower friction. Ice and steel easily slide past each other. For this reason, they have a low coefficient of friction.

Consider This

Plastics are said to have a low coefficient of friction. Why do you think that is? What are some positives and negatives of using plastic?

WORD WISE

A coefficient is a number that appears before and multiplies a variable, which is a symbol used to represent a value.

Faults with Too Much Friction

While friction is often helpful, it can also cause trouble. Too much can be harmful for machinery. Many parts in the **engines** of cars experience a lot of friction. Over time, this can cause those parts to wear down and break. Lubricants such as oil and grease are applied to these moving parts to decrease the amount of friction and help solve the problem.

It's important to put lubricant such as oil on a bike chain to prevent friction from wearing and tearing parts that make the bike move forward.

Sports cars often have a streamlined design to help them travel faster by reducing air resistance.

Objects also move through air. The air pushes against the object as it moves, slowing it down. This type of friction is known as air resistance. The bodies of birds are naturally round and smooth. This decreases air resistance, helping them fly. Airplanes, rockets, and high-speed trains are designed with the same kinds of rounded shapes. Creating an object with a rounded shape to help it move through air or water is called streamlining.

WORD WISE

An engine is a machine that changes energy from one source, such as heat from burning fuel, into mechanical motion.

Faults with Too Little Friction

Just as too much friction can be a problem, too little friction can also cause issues. For an airplane to land safely, the pilot sets the plane's flaps in the correct position. Air pushing against the flaps, or air resistance, allows the plane to slow down. Without the flaps, the plane would land at a higher speed and need a much longer runway to slow down and stop.

Car tires have treads to increase the amount of friction between the tires and the road. If the treads are worn down, this creates too little friction. The car may slip and spin out of control on wet or icy roads. Race cars can be more likely to slide off the track towards the end of a run because the tread on their tires gets worn down, decreasing their friction with the road.

A lack of friction can cause a car to lose "grip" with the road and to slide. That's why it's important to drive slowly in snowy and icy conditions.

compareandcontrast

A seaplane is a plane that can take off and land on water. How do you think friction is the same and different for a seaplane versus a commercial airplane?

A lack of friction from air or water could be problematic for a seaplane. Seaplanes have floats that create friction with water to slow down.

Natural Friction Forces

Animals also use friction in many ways. Snails and snakes creep and slither along the ground. Rocks and sticks may make it tough for them to travel. Snails have solved this. They produce their own natural lubricant, a slippery substance called mucus. The mucus allows the snail to glide wherever it needs to go. Snakes use the friction created by their scales to slither through their surroundings.

Snakes' scales provide friction with the ground. This allows them to move in certain directions.

Tunas' bodies are streamlined because they are torpedo shaped. They also have fins that fit into grooves on their bodies to further prevent water resistance.

Animals that live in the water deal with water resistance. Many fish and other underwater creatures have rounded heads and bodies. Others have long, slim heads and bodies. A tuna is one example of an underwater creature with a streamlined body, allowing it to move quickly and easily through the water.

Consider This

Can you think of other water creatures with a streamlined shape that helps reduce water resistance?

More Fun Friction Tricks

The force of friction can cause much more than a change in speed. The friction caused by two objects rubbing or sliding can cause them to become warm and even hot. By rubbing two sticks together, it is possible to make enough heat to start a fire.

Making a fire using friction can warm you up twice—first through the effort of making the fire, and then after the fire is lit!

Usually only male crickets chirp. They use a structure called a scraper that is found on top of their wings.

Quite often, chunks of rock from space enter Earth's **atmosphere**. These meteors travel so fast that the friction between the meteor and the atmosphere produces an extreme amount of heat. As a result, the surface of the meteor heats up and burns. We see this as a light in the night sky known as a shooting star.

Crickets create a familiar chirping sound through an action called stridulation. This involves rubbing body parts together to create sound through friction.

WORD WISE

An atmosphere is the mixture of gases that surround a planet.

Finding Friction Everywhere!

Our lives wouldn't be the same without friction. In our neighborhoods and communities, people of all ages use friction for their work and play. It's a force that cannot be seen, but its effects can be felt and observed. It will always be with us.

Special tires called snow tires have been invented to use on vehicles. These have larger and deeper grooves to create more friction between the tires and a snowy or icy road.

Many common lubricants to prevent friction from wearing and tearing spacecraft parts evaporate in space. Scientists are inventing new lubricants that can withstand space conditions.

Today and in the years to come, engineers will continue to learn about friction. They will create and improve products that use the helpful effects of friction, such as brakes, tires, and the soles of shoes. They will also continue to invent solutions to issues caused by too much or too little friction.

Consider This

Unlike on Earth, there is basically no friction in space. What challenges do you think this might cause for scientists studying space?

Glossary

concrete: A mix of water, stones, sand, and a soft gray powder that becomes very hard and strong when it dries.

design: The pattern or shape of something. Also, to create the pattern or shape of something.

engineer: A person who designs and builds complicated products, machines, systems, or structures.

evaporate: To change from a liquid to a gas.

extreme: To a very great degree.

lubricant: A substance used to decrease friction between surfaces that are in contact.

meteor: A piece of rock or metal that falls from space into Earth's atmosphere, where the heat of friction causes it to burn and glow brightly in the sky.

microscope: An instrument used for producing enlarged images of very small objects so that they can be seen clearly.

molecule: A very small piece of matter.

mucus: A slippery substance produced by animals to moisten and protect body passages that are in contact with the outside world.

stridulation: A process in which insects make a shrill sound by rubbing two body parts together.

symbol: A picture, shape, or object that stands for something else.

thrilling: A word to describe something that is very exciting.

torpedo: A long, thin bomb. It is often fired from a ship to destroy another ship and moves underwater.

tread: The ridges or grooves on the surface of a tire or on the bottom of a shoe.

vehicle: An object used for carrying or transporting people or goods, such as a car, truck, or airplane.

water resistance: Friction that affects objects moving through water.

For More Information

Books

Brundle, Joanna. *Forces*. New York, NY: KidHaven Publishing, 2020.

Claybourne, Anna. *Recreate Discoveries About Forces*. New York, NY: Crabtree Publishing, 2019.

Donnelly, Rebecca. *How Slippery Is a Banana Peel?* New York, NY: Henry Holt and Company, 2021.

Websites

Britannica Kids: Friction
kids.britannica.com/kids/article/friction/442495
Read on about what friction is, where it is found, and how it can create problems and solutions.

Physics for Kids: Friction
www.ducksters.com/science/friction.php
Learn more about how friction works and how it is related to energy here! Then test what you learned with a short quiz.

Why Is Ice Slippery?
www.wonderopolis.org/wonder/why-is-ice-slippery
Find out more about how a lack of friction makes ice slippery here.

Index